SpringerBriefs in Environmental Science

SpringerBriefs in Environmental Science present concise summaries of cutting-edge research and practical applications across a wide spectrum of environmental fields, with fast turnaround time to publication. Featuring compact volumes of 50 to 125 pages, the series covers a range of content from professional to academic. Monographs of new material are considered for the SpringerBriefs in Environmental Science series.

Typical topics might include: a timely report of state-of-the-art analytical techniques, a bridge between new research results, as published in journal articles and a contextual literature review, a snapshot of a hot or emerging topic, an in-depth case study or technical example, a presentation of core concepts that students must understand in order to make independent contributions, best practices or protocols to be followed, a series of short case studies/debates highlighting a specific angle.

SpringerBriefs in Environmental Science allow authors to present their ideas and readers to absorb them with minimal time investment. Both solicited and unsolicited manuscripts are considered for publication.

More information about this series at http://www.springer.com/series/8868

Fernando Ramírez · Josefina Santana

Environmental Education and Ecotourism

Springer

Fernando Ramírez
Independent Researcher
Bogotá, Colombia

Josefina Santana
School of Communication
Universidad Panamericana
Guadalajara, Jalisco, Mexico

ISSN 2191-5547 ISSN 2191-5555 (electronic)
SpringerBriefs in Environmental Science
ISBN 978-3-030-01967-9 ISBN 978-3-030-01968-6 (eBook)
https://doi.org/10.1007/978-3-030-01968-6

Library of Congress Control Number: 2018958360

This Springer imprint is published by the registered company Springer Nature Switzerland AG
The registered company address is: Gewerbestrasse 11, 6330 Cham, Switzerland

The first author dedicates this book to his Mother (Natalia) and Father (Fernando) and to L. Marien.

Preface

Educators are calling for a move from traditional, classroom-based teacher-centred practices to innovative learner-centred approaches. This is not just a fashion; rather, the call obeys an understanding that, for learning to be meaningful, it must involve the student holistically. Hands-on educational practices lead to deeper understanding and long-term learning.

On the other hand, fostering environmental awareness among newer generations is crucial. The planet strains to accommodate the growing human population and its activities. Thus, environmental education is imperative and it is not enough to discuss the subject in the classroom. The learners must go out and experience nature in a thoughtful, intentional manner.

Ecotourism is a sound and environmentally friendly field that currently offers new perspectives to the tourism industry. The connection between ecotourism and education is key for biodiversity conservation and caring for the environment.

The purpose of this Springer Brief is twofold. Firstly, it seeks to promote environmental education by offering guidelines in ecotourism as a school subject. Second, and more importantly, it seeks to raise awareness of the role we must play in preserving our world.

Bogotá, Colombia
Guadalajara, Mexico
April 2018

Fernando Ramírez
Josefina Santana

Contents

Abstract

Ecotourism is an important field of tourism that has bridged environmentally friendly practices and the leisure industry worldwide. Today, ecotourism has become an important revenue sector for stakeholders and communities. It has provided numerous solutions as a suitable tourism type, but has also caused environmental impacts. Recently, ecotourism has become important in developing countries, offering a sustainable practice, i.e. ecologically friendly minimizing the impacts caused by other tourism practices. Education is a key aspect for ecotourism because it has provided a dialogue between communities, stakeholders, researchers, citizens and ecotourists. Environmental education has provided a framework for any ecotourism practice, because it encourages the basic know-how and principles for a environmentally friendly practice. Environmental education is a core aspect in ecotourism because it can provide the necessary momentum to encourage the common citizen to become aware of ecotourism practices and perform responsible ecotourism. This Springer Brief focuses on environmental education and ecotourism and analyses biodiversity conservation, philosophy, environmental impacts, indigenous ecotourism and wetland ecotourism. Also, this book is supported by recent publications from the authors.

Chapter 1
Introduction

Worldwide ecotourism has become a new perspective for integrating nature, culture, and the tourism sector. Though ecotourism has been touted as a "green"—that is, ecologically sound—form of tourism, it can paradoxically lead to the degradation of the environment.

A case in point is that of the "Pueblos Mágicos" of Mexico. These "Magic towns" are little-explored culturally and aesthetically rich sites. They have been selected by the Mexican Secretariat of Tourism based on several criteria—including natural beauty—and they receive special promotion and funding. Currently, there are just over one hundred such towns around the country, including the town of Tequila in Jalisco, and its surrounding area (Fig. 1.1).

Being named a "Pueblo Mágico" brings the region needed funds, as well as additional tourism. Unfortunately, the site is not always equipped to provide for the influx of visitors. Hotels, restaurants, and other businesses spring up rapidly to cater to tourists, and basic infrastructure such as waste management and traffic control are strained. In consequence, the site ends up losing some of its original magic.

This example illustrates the paradox of ecotourism. The economic benefits include jobs for the local population, reducing the need to emigrate. The downside is the loss of traditional ways of life, and the destruction of some of the natural beauty of the site.

This book posits that environmental education may be useful in balancing economic benefits and sustainability in ecotourism.

1.1 Environmental Education

Environmental education can be understood as the process of gaining awareness of the world around us, with its natural beauty and resources. Ideally, this understanding should not be only theoretical. It should lead the individual to value nature and to seek to preserve it.

F. Ramírez and J. Santana, *Environmental Education and Ecotourism*, SpringerBriefs in Environmental Science, https://doi.org/10.1007/978-3-030-01968-6_1

Fig. 1.1 Round pyramids at Guachimontones, near the magic town of Tequila, Jalisco. Photo by Rafael Santana. Reproduced with permission

For environmental education to be effective in the context of ecotourism, it must be directed to three distinct populations. The first is visitors to sites. They have a unique opportunity to discover in situ the beauty and diversity offered by the site. The second group is the local population, who must learn how to exploit their natural resources in ways that are both productive and sustainable. Finally, future generations should be taught the importance of conserving the environment and the best ways to do so.

The following sections will discuss the current state of affairs in education among these three groups.

1.2 Educating Visitors

Ceballos-Lascurain (1988), who coined the term ecotourism, insisted that one of its specific goals was to *study* the scenery, the culture, the plants and the animals of the region. However, it is not clear if participating in ecotourism can lead to increased conservation practices. Sander (2012), for example, mentions that most participants in ecotourism are already environmentally aware. Beaumont (2001) however, found that ecotourism serves to reinforce this awareness, and that the greatest gains in awareness are obtained by ecotourists with the least environmental knowledge.

It seems that just having information about the site available to visitors has an impact on their knowledge, though this does not necessarily lead to changes in behaviors (Sander 2012). Ideally, environmental education should inspire the learner to take concrete actions toward conservation.

1.3 Educating the Local Population

Another goal of ecotourism should be the education of the local population. They need to know, not only how to conserve their natural resources, but how to transmit knowledge of these resources to visitors. Sander (2012) mentions a partnership between a university in the United States and the Kayapó indigenous group in Brazil. The Kayapó functioned as guides and teachers to students from the university, but ended up learning from the students, as well. The authors of the study, Zanotti and Chernela, cited in Sander (2012) mention four benefits of the programme for the local people. These are increased revenue; "the opportunity to control the presentation of knowledge about themselves and their territories" (p. 394); the chance to practice a new language; and the possibility of expanding their support network.

Any scheme to educate the local population must consider the perspective of Indigenous Knowledge (IK). In Grenier's 1998 definition, cited in Sillitoe et al. (2005), this is "The unique, traditional, local knowledge existing within and developed around specific conditions of women and men indigenous to a particular geographic region" (p. 1). IK has been applied to many domains of study, but it is of particular value in ecotourism, as the local population not only has the greatest knowledge of the region under study, but is also the most important group of stakeholders.

1.4 Educating Future Generations

In developing countries, studies into environmental education frequently start from the stance of Social Representation Theory (SRT). D'Amato-Herrera (2012), for example, cites important studies in Argentina, Brazil, Colombia, and Mexico which use SRT to establish what learners believe about the world around us.

Social Representations are beliefs formed from a mixture of our culture and our society, incorporated into our commonsense knowledge of the world. These social representations (SR) then lead us to create our explanations of reality (Cuevas-Cajiga 2012). Taking SR as an object of study leads to the construction of knowledge which will become the basis of action and interpretation of a community (D'Amato-Herrera 2012).

Meira Cartea (2013) states that "It is from these representations built individually and socially that we can give sense and meaning to the world around us, share it with others and guide our attitudes and behavior with respect to it" (p. 32). With regards to environmental education, for example, SR are responsible for our beliefs about natural resources: are they to be exploited for our benefit? Or are they to be cherished and preserved?

Studies into the effectiveness of environmental education in schools have shown mixed results. Among university students, Correa-López and Ortiz-Espejel (2012) found that visits to actual sites, and volunteer work carried out among the local community were the most effective means to make an impact on learners. Otherwise, the knowledge acquired in their classes is merely theoretical.

Calixto-Flores (2012) studied middle school students (12–15 year-olds) and found that their SR tend to equate the environment exclusively with nature without considering social or cultural dimensions. This is true among teachers, as well.

Terrón-Amigón (2012) found that among teachers, knowledge was not uniform. The SR she found in teachers were classified as simple, globalizing, anthropocentric, integral, or critical. It is the latter two that lead to changes in perceptions of the environment, but these SR were found in only a few of the teachers interviewed. The author found that most teachers associate environmental education exclusively with its physical facet, without regard for either social or cultural aspects.

Thus, even though class time is devoted to environmental issues, these are usually seen in a superficial and theoretical way, and do not lead to changes in perception or actions among the learners.

As Saylan and Blumstein (2011) express "Environmental education must go beyond recycling programmes and teach us how to achieve measurable and increasing impact by reducing our human footprint substantially in a lifelong endeavor… Environmental education must clearly illustrate that there is only one earth, and we're all on it together" (p. 29).

This book hopes to contribute to this end.

References

Beaumont N (2001) Ecotourism and the conservation ethic: recruiting the uninitiated or preaching to the converted? J Sustain Tourism 9(4):317–338. https://doi.org/10.1080/09669580108667405

Calixto-Flores R (2012) Miradas de los estudiantes de educación secundaria sobre el medio ambiente. In: Calixto R (ed) En la búsqueda de los sentidos y significados de la educación ambiental. Universidad Pedagógica Nacional, Mexico City, pp 107–128

Ceballos-Lascurain H (1988) The future of ecotourism. Mexico J 13–14

Correa-López MI, Ortiz-Espejel B (2012) Transformación de las RS a través de la educación ambiental en el nivel universitario. In: Calixto R (ed) En la búsqueda de los sentidos y significados de la educación ambiental. Universidad Pedagógica Nacional, Mexico City, pp 191–214

Cuevas-Cajiga YM (2012) Apuntes sobre la teoría de representaciones sociales. In: Calixto R (ed) En la búsqueda de los sentidos y significados de la educación ambiental. Universidad Pedagógica Nacional, Mexico City, pp 21–44

D'Amato-Herrera G (2012) Las representaciones sociales y la psicología ambiental como dinamizadores de la educación ambiental. In: Calixto R (ed) En la búsqueda de los sentidos y significados de la educación ambiental. Universidad Pedagógica Nacional, Mexico City, pp 45–56

Meira Cartea PA (2013) Problemas ambientales globales y educación ambiental: Una aproximación desde las representaciones sociales del cambio climático. Revista Integra Educativa 6(3):29–64

Sander B (2012) The importance of education in ecotourism ventures: lessons from Rara Avis ecolodge, Costa Rica. Int J Sustain Soc 4:389–404. https://doi.org/10.1504/IJSSOC.2012.049408

Saylan C, Blumstein D (2011) The failure of environmental education (and how we can fix it). University of California Press, Los Angeles

Sillitoe P, Dixon P, Barr J (2005) Indigenous knowledge inquiries: a methodologies manual for development. Practical Action Publishing, Rugby
Terrón-Amigón E (2012) Horizontes de la educación ambiental. Un estudio de representaciones sociales. In: Calixto R (ed) En la búsqueda de los sentidos y significados de la educación ambiental. Universidad Pedagógica Nacional, Mexico City, pp 129–148

Chapter 2
Environmental Education and Biodiversity Conservation

The term *biodiversity* refers to the wealth of life on Earth. The United Nations Convention on Biological Diversity (CBD) defines it thus: "'Biological diversity' means the variability among living organisms from all sources including, inter alia, terrestrial, marine and other aquatic ecosystems and the ecological complexes of which they are part; this includes diversity within species, between species and of ecosystems" (CBD, n/d) (Fig. 2.1).

The 1970s brought about an increasing awareness of how humankind was causing irrevocable damage to these ecosystems. Clearing of forests for agricultural purposes, overfishing, dam construction, and other human practices all take their toll on the environment. These practices, designed to make human life more practical,

Fig. 2.1 The biodiversity rich cloud forest at La Aguadita, near Fusagasugá, Cundinamarca State, Colombia. **a**, **b** Mountain view, **c**, **d** detail of forest with arborescent ferns. Photos by Fernando Ramírez. Reproduced with permission

F. Ramírez and J. Santana, *Environmental Education and Ecotourism*, SpringerBriefs in Environmental Science, https://doi.org/10.1007/978-3-030-01968-6_2

threaten, not only diverse animal and plant species, but humankind as well. It became necessary to establish plans and protocols to protect the environment. At the same time, it was necessary to consider that people have a right to earn their livelihood. Reconciling these two seemingly contradictory needs gave rise to the notion of sustainable development—preserving nature while accepting that people have the right to development, understood in the economic sense (Bonnet 1999).

Because environmental problems tend to be based on lack of knowledge and education (Valderrama-Hernandez et al. 2017), promoting biodiversity conservation is an important goal of education worldwide (Ramadoss and Poyya Moli 2011). The United Nations has taken on a leading role in promoting environmental education, through diverse initiatives. One of these is the UN Conference of Environment and Development in Rio de Janeiro in 1992. Three programs were proposed: education for sustainable development, increased public awareness, and training of selected people to monitor and manage the environment (Kassas 2002).

Educators were urged to include environmental education in their curricula, not only in biology and natural science classes, but across the spectrum, as a topic to be discussed in social sciences, and other school subjects. Kassas (2002) situates education for biodiversity conservation on five "pivots". These are:

- Scope. Environmental education has been a complex topic because it is difficult to agree on the focus. Advocacy groups such as Greenpeace, scientists, and ecotourists all care about nature, but they view and value it in different ways.
- Perspectives. Here, the author mentions three different types of relationships: that among different species, a personal relationship with nature, and a supportive society that seeks to protect nature.
- Goals. These are aligned with the four "pillars" which should support environmental education, as established by the United Nations Educational, Science and Culture Organization (UNESCO) in its goals for sustainable development. The first of these pillars is biophysical systems; that is, the interrelatedness of all living things. The second pillar is economic systems, which include the local people's ability to sustain themselves. The third is social systems, and the final pillar is political systems, which include the local people's right to a say in regards to environmental issues (Fien 2004).
- Themes. These are the issues which are of particular interest to a community or region, or at a specific point in time.
- Assimilation. Kassas (2002) here talks about the need to evaluate how each actor has taken on their corresponding roles.

The rest of this chapter will be organized according to these five pivots.

2.1 Scope

Though most people would agree that conservation of nature is a worthy goal, not everyone agrees on how to accomplish that goal. Efforts in the past decade have focused on sustainable development. UNESCO (2017), in its Agenda 2030, has established this as its "first and foremost" objective "shifting the world on to a sustainable and resilient path" (p. 3).

However, some authors (Ferreira 2002, for example), question whether sustainable development and nature conservancy might not be mutually exclusive. Bonnet (1999) for his part, questions: what is to be sustained? Is it nature? Or economic growth? Current lifestyles? It cannot be all of these. The author mentions that most current programmes view environmental education from the viewpoints of science and geography. Meanwhile, other equally important aspects, such as the social, ethical, economic, aesthetic, political, cultural, and spiritual dimensions of ecology are forgotten.

2.2 Perspectives

Costa Rica has established itself in the forefront for biodiversity conservation. It is an environmentally rich and diverse country, and one of its main sources of income is ecotourism. In 2005, the national government implemented a national strategy for environmental education, intended to "increase awareness, knowledge, values, skills, and goodwill to act for the natural, cultural, and spiritual resources conservation" (Jimenez et al. 2017, p. 223). These authors studied different local programmes focused on Communication, Education, Public Awareness, and Participation Strategies. They found that most programmes fall into two types, categorized by the authors as *vision of nature protection* and *vision of sustainability*. The first of these is geared towards schoolchildren and teaches them basic biological knowledge. The second type of program focuses on adults, and focuses on "community-based management of natural resource projects" (p. 235).

The programmes are delivered at schools or at environmental educations centers, such as botanical gardens or wildlife preserves.

2.3 Goals

Biodiversity conservations strategies which focus efforts on developing countries are unfair, as more developed countries are the greatest consumers of natural resources. The principle of subsidiarity establishes that those who have greater power also have greater responsibility (Francis 2015).

Any environmental protection programme needs to consider that biodiversity includes not only plant and animal species, but humankind as well. Deforestation, soil erosion, water pollution, and fish depletion are all grave environmental problems which, in the long run, will affect the entire planet. However, first in the line of fire, as it were, are the communities which are closest to natural resources, and whose livelihood depends on them. Caring for natural resources is important, but human lives are important, as well. Thus, true ecological programmes should also be social programmes (Francis 2015).

2.4 Themes

Climate change is a global concern, as its consequences affect the whole planet (Ramírez and Kallarackal 2015, 2018). Other issues, such as the protection of the Monarch Butterfly Biosphere Reserve, in the Michoacán region of Mexico are of local concern. The local community of Michoacán relies greatly on ecotourism for its income, but deforestation and illegal logging threaten the Reserve, where the Monarch butterflies hibernate (Manzo-Delgado et al. 2014). Thus, educational efforts in this region are focused on learning about the Reserve and its importance, and efforts to protect it.

Another example of education of the local population also takes place in Michoacán. Efforts to protect the critically endangered *achoque* amphibian [*Ambystoma dumerilii* (Dugès, 1870)] are carried out by nuns at a local convent (Frías-Alvarez et al. 2010). The *achoque* is used as a source of food and medicine by the indigenous population. For centuries, the convent has produced a cough syrup based on the amphibian's skin, but water pollution in the *achoque's* habitat was rapidly decreasing the population of the amphibian. The nuns learned, with the help of a biologist monk, how to breed the amphibians in captivity, and how to care for them. Currently, their convent functions as a research unit focused on the conservation of the species, but providing also a source of income for the local population. The nuns participate in research conferences and have written a book on the topic (Pérez-Saldaña et al. 2006).

2.5 Assimilation

As established by the United Nations Convention on Biological Diversity, in its strategic plans for biodiversity 2011–2020, local communities and indigenous populations are to be included in the development of these plans (CBD n/d). In the same vein, environmental education programmes (EEP) should not be limited to abstract content studied in classrooms; those first affected by environmental issues should be targeted as well.

One such programme has been carried out in Nepal (Shrestha 2015). The study looked at the effect of EEPs on wildlife conservation on behaviours and attitudes. It found that local people who had attended the EEPs expressed a more positive attitude toward conservation goals than those who had not attended. This was seen also in a decrease in poaching incidents in the region under study.

Many authors agree that, for environmental education to be effective, it must change behaviours and attitudes, and develop values (Arslan 2012; Zakharova et al.

2015, among others). Where there is less agreement is on what and how it is to be taught. As this chapter shows, environmental education is a complex issue which seems to raise more questions than can be answered simply. What can be said, however, is that it is an issue which must be addressed, not only in the classroom, but in the field, and in the family, and in religious and political institutions, and in firms. In the care of the planet we are all stakeholders, and we must all work towards protecting our common home.

References

Arslan S (2012) The influence of environment education on critical thinking and environmental attitude. Proc Soc Behav Sci 55:902–909

Bonnet M (1999) Education for sustainable development: a coherent philosophy for environmental education? Camb J Educ 29(3):313–324

Ferreira JG (2002) Biodiversity and environmental education: a contradiction? Koers J 67(3):259–269

Fien J (2004) Education for sustainability. In: Gilbert R (ed) Studying society and environment: a guide for teachers. Cengage Learning Australia, Melbourne

Francis (2015) Encyclical letter Laudato si' of the Holy Father Francis, 1st ed. Retrieved from http://w2.vatican.va/content/francesco/en/encyclicals/documents/papa-francesco_20150524_enciclica-laudato-si.html

Frías-Alvarez P, Zúñiga-Vega JJ, Flores-Villela O (2010) A general assessment of the conservation status and decline trends of Mexican amphibians. Biodivers Conserv 19(13):3699–3742. https://doi.org/10.1007/s10531-010-9923-9

Jimenez A, Monroe MC, Zamora N, Benayas J (2017) Trends in environmental education for biodiversity conservation in Costa Rica. Environ Dev Sustain 19:221–238. https://doi.org/10.1007/s10668-015-9734-y

Kassas M (2002) Environmental education: biodiversity. Environmentalist 22:345–351

Manzo-Delgado L, López-García J, Alcántara-Ayala I (2014) Role of forest conservation in lessening land degradation in a temperate region: the Monarch butterfly biosphere reserve, Mexico. J Environ Man 138:55–66

Pérez-Saldaña MC, Gutiérrez Acosta M, Morales Francisco O, Pérez-Saldaña J (2006) Experiencias de cultivo de achoque (Ambystoma dumerilii) en cautiverio. Monasterio de Dominicas de Orden Predicadores María Inmaculada de la Salud AR. Pátzcuaro, Michoacán

Ramadoss A, Poyya Moli G (2011) Biodiversity conservation through environmental education for sustainable development—a case study from Puducherry, India. Int Eletron J Environ Educ 1:97–111. https://doi.org/10.18497/IEJEE-GREEN.99495

Ramírez F, Kallarackal J (2015) Responses of fruit trees to global climate change. Springer, Switzerland

Ramírez F, Kallarackal J (2018) Tree pollination under global climate change. Springer, Switzerland

Shrestha S (2015) The role of environmental education for biodiversity conservation: a case study in the protected areas of Nepal. Unpublished doctoral thesis, Arizona State University

UNESCO (2017) UNESCO moving forward the 2030 agenda for sustainable development. Retrieved from http://unesdoc.unesco.org/images/0024/002477/247785e.pdf

Valderrama-Hernandez R, Alcantara L, Limon D (2017) The complexity of environmental education: teaching ideas and strategies from teachers. Proc Soc Behav Sci 237:968–974

Zakharova E, Liga M, Sergeev D (2015) Constructing philosophy of environmental education in contemporary Russia. Proc Soc Behav Sci 214:1181–1185

Chapter 3
Relevance of Ecotourism

Today's globalized world has an overproduction of goods and services. The tourism sector offers multiple destinations, tours and attractions. In consequence, this has caused major environmental problems i.e., sewage, garbage, land use, reduction in forest cover, coastal modification, habitat loss for many species, etc. Today, ecotourism has provided a better way to manage resources and offers an environmentally-based ideology. Also, it proposes important links between conservation and restoration. Ecotourism can be applied successfully for ecosystem species and conservation i.e. of critically endangered biodiversity (Bookbinder et al. 1998; Ghosh and Ghosh 2018; Wardle et al. 2018). Through economic incentives and revenues, ecotourism has provided local guardianship by communities or villagers, while encouraging effective resource management (Bookbinder et al. 1998; Eshoo et al. 2018). This tourism field has been important in ecosystem conservation worldwide and particularly in wetlands (Ramírez and Fennell 2014; Ramírez and Santana 2018) (Fig. 3.1), deserts, rainforests (Fig. 3.2), and grasslands among other ecosystems.

Education is a fundamental aspect that needs to be analyzed in any ecotourism discussion. Education has implications that permeate the ecotourism researcher, ecotourist, target communities, etc. This is because it teaches how to interact with nature in harmony and consciously. Education plays an important role in biodiversity conservation by defining the basic principles that rule the practice of ecotourism. Ecotourism plays an important role and has great potential in environmental education and sustainability (Kimmel 1999; Gilbert 2003; de Lima et al. 2018; Ramírez and Santana 2018). The relevance of education in ecotourism is an aspect that has been examined in ecotourism ventures (Sander 2012; Mendoza-Ramos and Prideaux 2018), wetland settings (Ramírez and Santana 2018), as a key aspect for the development of tourism (Stergiou et al. 2008), core to its fundamental aspects, definitions and philosophy.

Ecotourism is an important field for conservation at both national and international levels. This field has generated important economic gains, and has also provided ecological sustainable development in several places around the world (Courvisanos and Jain 2006; Howitt and Mason 2018). Also, it has attracted the interest of entrepreneurs, local communities, governments, non-governmental institutions

F. Ramírez and J. Santana, *Environmental Education and Ecotourism*, SpringerBriefs in Environmental Science, https://doi.org/10.1007/978-3-030-01968-6_3

Fig. 3.1 Wetland ecosystems. Loretoyacu River in the Colombian Amazon (top photo) and The Cuicocha Lagoon, Ecuador (bottom photo). Photos by Fernando Ramírez. Reproduced with permission

and other stakeholders from the tourism sector. Ecotourism as a field is well established in Australia, North America, and Europe. It is currently under development in South America, parts of Central America, Asia and Oceania. Developing countries are increasing programmes linking biodiversity conservation. This provides strong thrust for caring for the environment. In Central America, Costa Rica provides a good example of a successful ecotourism programme which has gained an international reputation as a leader in environmental conservation. This is key for developing a

Fig. 3.2 Amazon rainforest near Puerto Nariño, Colombia. **a**, **b** Rainforest and **c** the colorful Guaraná (*Paullinia cupana* Mart.) fruit. Photos by Fernando Ramírez. Reproduced with permission

reliable and profitable business (Courvisanos and Jain 2006). Ecotourism, has been considered an important field for economic development and conservation (Honey 2008). Developing countries have been offering and promoting several ecotourism brands. This includes countries such as Colombia, Bolivia, Dominica, Belize, Mongolia, Vietnam, Argentina, Bhutan, Fiji, Namibia, Indonesia, Peru, Senegal, Madagascar, Ecuador, Cambodia, Thailand, Uganda, Australia, New Zealand, and Canada, which are actively offering and providing interesting ecotourism destinations (Honey 2008).

Ecotourism has been practiced in environmentally pristine and protected areas that are considered unique, characterized by their ecological interest and cultural importance (Wearing and Neil 1999; Martin et al. 2018), but it can also be practiced in human-influenced areas (Fennell 2013). Today, these areas have been established by biodiversity conservation purposes that seek to preserve ecosystems (Wearing and Neil 1999; Schulze et al. 2018), species, habitats, populations and communities. Ecotourism is an important field for establishing reliable and effective economic use of protected areas (Higham 2007). Numerous regions and people have developed ecotourism as independent ideas connecting environmental awareness and ecology (Fennell 1998). Worldwide the international conservation sector has generated projects, programmes, and research related initiatives, that have been generating nature-based tours encouraging ecotourism (Honey 2008). Moreover, international organizations and aid agencies, have pushed forward nature related aspects such as local revenue generation, biodiversity aspects, sustainable rural development, institutional organization, and consistent infrastructure by funding projects related to the tourism sector, that are of the ecotourism nature (Honey 2008). This can be applied to different cultural contexts and places worldwide. Many developing countries are near the Equatorial zone, characterized by a high diversity of species and ecosystems. This is one of the major aspects attracting thousands of visitors from different latitudes to visit tropical countries. For example, the Colombian Amazon provides one of the most spectacular places for ecotourism. Leticia, the capital of the Amazon State is a city of thousands of inhabitants. This city, is bordered by the Amazon River (Fig. 3.3), which forms a series of oxbow lakes, remnants of the main channel. One of these lakes, Yahuarcaca, is home to numerous species, from the largest scale fish, the Pirarucú (*Aramaima gigas*), to river dolphins, manatees, eagles, and turtles (Fig. 3.4). Flowing into the seasonal Yahuarcaca Lake is the blackwater *terra firme* Yahuarcaca stream (Fig. 3.5). In Amazonian streams, tree canopies prevent light from reaching the water surface directly, nutrient salts are scarce, and aquatic plant life is virtually non-existent. Furthermore, food webs are highly influenced by resources coming from the nearby terrestrial environment (Lowe-McConnell 1987). The Yahuarcaca stream is home to 171 fish species which have numerous dietary-morphological specializations (Ramírez et al. 2015).

Due to the environmental relevance that ecotourism has gained, many countries are including legislations that support ecotourism planning, development and enhancement. Government policy has been used to push forward the necessary environmental regulations, industry related aspects minimizing the negative impacts, providing and proposing sound environmental practices for ecotourism (Wearing and Neil 1999). Governmental policy planning has been an important tool for developing and building reliable ecotourism projects (Wearing and Neil 1999). Nowadays, more governments are using ecotourism as a key source to generate revenues (Wearing and Neil 1999). This is particularly the case of countries in Central America such as Costa Rica and Mexico. Legislation has to focus on aspects that regulate the ecotourism industry such as carrying capacity, revenue collection, mechanism of control for tour operators, programme planning, infrastructure, communications media and international cooperation. Legislators are at the heart of decision making. Although developed countries have a consistent and robust framework for policy and integration with ecotourism (see, for example, the case of the U.S.A. or Australia), developing countries tend to have serious cases of corruption which negatively impact any ecotourism programme, as allocated funds are often misused or retained by politicians that exert a powerful influence. Thus, individual interest overruns community interest leading to corruption and environmental crises. According to Wearing and Neil (1999), an ecotourism plan should ideally encompass the following steps:

- research preparation
- goal identification
- examination through surveys
- discussion

Fig. 3.3 Amazon River near Leticia, Colombia. Photos by Fernando Ramírez. Reproduced with Permission

- policies and planning construction
- recommendations
- implementation and monitoring

Effective planning and legislations are key aspects for ecotourism development. Strong legislations at local, national and international levels are also fundamental aspects that are required for an operational ecotourism sector. Moreover, cooperative aid between the parts involved provides an important point of departure for effective success (Wearing and Neil 1999).

Philosophically, ecotourism is a tourism field distinguished by its environmental concern and appreciation for nature. Caring for the environment has long been a major

Fig. 3.4 Birds from the Amazon region of Colombia. Photo by Fernando Ramírez

Fig. 3.5 The Yahuarcaca stream near Leticia, Colombia **a** detail and **b**, **c** characin fish species. Photos by Fernando Ramírez. Reproduced with permission

premise that concerns nations and governments, as well as the local citizen. Today's globalized world proposes numerous links between cultures and, thus, a plethora of possibilities for environmental change and consciousness. Well-founded ecotourism ideas provide a framework for differentiating ecotourism practices from pseudo- or non-ecotourism based ideologies. Purely economic and extractive practices are often referred to as having a different objective from that of ecotourism. Sound ecotourism practices require ideologies based on what is meant by ecotourism. Over the decades, since the foundation of ecotourism, there has been a lot of debate over philosophical and fundamental aspects of ecotourism. Currently, there is a need to reshape the field of ecotourism in face of its fundamental philosophical basis. Few investigations have delved into the philosophy of ecotourism and more research is clearly needed to better understand its profound meaning and links to the environment, human beings, tourism and other fields of knowledge.

References

Bookbinder MP, Dinerstein E, Rijal A, Cauley H, Rajouria A (1998) Ecotourism's support of biodiversity conservation. Conserv Biol 12:1399–1404

Courvisanos J, Jain AA (2006) A framework for sustainable ecotourism: application to Costa Rica. Tourism Hospit Plann Dev 3:131–142

de Lima L, Dutra Neto L, de Holanda L (2018) programa de Ecoturismo e Educação Ambiental do Projeto TAMAR - Fernando de Noronha: uma análise de 2013 a 2016. Rev Bras Ecoturismo 11:47–60

Eshoo PF, Johnson A, Duangdala S, Hansel T (2018) Design, monitoring and evaluation of a direct payments approach for an ecotourism strategy to reduce illegal hunting and trade of wildlife in Lao PDR. PLoS ONE 13:e0186133. https://doi.org/10.1371/journal.pone.0186133

Fennell DA (1998) Ecotourism in Canada. Ann Tourism Res 25:231–234

Fennell DA (2013) Contesting the zoo as a setting for ecotourism, and the design of a first principle. J Ecotourism 12:1–14

Ghosh P, Ghosh A (2018) Is ecotourism a panacea? Political ecology perspectives from the Sundarban Biosphere Reserve, India. GeoJournal 1–22. https://doi.org/10.1007/s10708-018-9862-7

Gilbert R (2003) Ecotourism and education for sustainability: a critical approach. Int Rev Environ Strateg 4:75–83

Higham J (2007) Critical issues in ecotourism: understanding a complex tourism phenomenon. Elsevier, Amsterdam

Honey M (2008) Ecotourism and sustainable development: who owns paradise? 2nd edn. Island Press, Washington, DC

Howitt J, Mason C (2018) Ecotourism and sustainable rural development in Pérez Zeledón, Costa Rica. J Rural Community Dev 13:67–84

IUCN (2008) What is a protected area? https://www.iucn.org/about/work/programmes/gpap_home/pas_gpap/

Kimmel JR (1999) Ecotourism as environmental learning. J Environ Educ 30:40–44. https://doi.org/10.1080/00958969909601869

Lowe-McConnell RH (1987) Ecological studies in tropical fish communities. Cambridge University Press, London

Martin A, Myers R, Dawson NM (2018) The park is ruining our livelihoods. We support the park! Unravelling the paradox of attitudes to protected areas. Hum Ecol 46:93–105. https://doi.org/10.1007/s10745-017-9941-2

Mendoza-Ramos A, Prideaux B (2018) Assessing ecotourism in an Indigenous community: using, testing and proving the wheel of empowerment framework as a measurement tool. J Sustain Tour 26:277–291. https://doi.org/10.1080/09669582.2017.1347176

Ramírez F, Fennell D (2014) A comprehensive framework for ecotourism and wetland restoration: the case of Bogotá, Colombia. J Ecotourism 13:128–151

Ramírez F, Santana JC (2018) Key lessons learned by teaching ecotourism to undergraduate students in Bogotá's urban wetlands. Appl Environ Educ Commun 1–18. https://doi.org/10.1080/1533015x.2018.1454359

Ramírez F, Davenport TL, Mojica JI (2015) Dietary–morphological relationships of nineteen fish species from an Amazonian terra firme blackwater stream in Colombia. Limnologica 52:89–102

Sander B (2012) The importance of education in ecotourism ventures: lessons from Rara Avis ecolodge, Costa Rica. Int J Sustain Soc 4:389. https://doi.org/10.1504/IJSSOC.2012.049408

Schulze K, Knights K, Coad L et al (2018) An assessment of threats to terrestrial protected areas. Conserv Lett e12435. https://doi.org/10.1111/conl.12435

Stergiou D, Airey D, Riley M (2008) Making sense of tourism teaching. Ann Tourism Res 35(3):631–649. https://doi.org/10.1016/j.annals.2008.03.001

UNEP (2015) Mapping the world's special places. http://www.unep-wcmc.org/featured-projects/mapping-the-worlds-special-places

Wardle C, Buckley R, Shakeela A, Castley JG (2018) Ecotourism's contributions to conservation: analysing patterns in published studies. J Ecotourism 1–31. https://doi.org/10.1080/14724049.2018.1424173

Wearing S, Neil J (1999) Ecotourism: impacts, potentials and possibilities. Reed Educational and Professional Publishing Ltd., Boston

Chapter 4
Education and Ecotourism

Early cultures in human history used nature as a source of food, tools, habitat, shelter, clothing, etc. After these basic needs for food, housing and clothing were fulfilled, humans acquired an interest in nature: in its intrinsic value. This interest became evident by early pictographic paintings on rocks at several locations worldwide and early art manifestations. Later on, cultures such as the Babylonians, Egyptians and Greeks applied observation to delve into the natural world. The observation of nature provided the first tool to establish a link between humans and nature. As production methods became more efficient (Arendt 2012), humans had more time for appreciating the natural world. Tourism as a field became an important part of the early link between the interest of studying cultures and nature. The interest of traveling to nature-based destinations became appealing for tourists who became a growing sector in the world's economy. This is the case of Villa de Leyva, Colombia, considered an interesting location for tourism and nature-based activities due to its Spanish historical background and nature-based destinations (Fig. 4.1). Other cases of nature-based destinations are Curaçao and Cartagena, Colombia (Figs. 4.2 and 4.3).

The tourism industry is currently an important economic sector within the world economy. This sector has generated large profits and services, but the tourism sector has positively and negatively influenced regions by physically transforming them (Fennell 2008). Tourism has been considered a reliable source for long-term regional development opportunities, but has also caused ecological damage and environmental issues, transforming regions completely (Fennell 2008). Moreover, mass tourism has been considered controversial because it dominates tourism within a region and by diverting revenues away from local to international destinations (Fennell 2008). Due to the multiple environmental problems and lack of care for the environment proposed by mass tourism, alternative forms of tourism have been generated. Alternative tourism focuses on minimising the negative effects generated by mass tourism (Wearing and Neil 1999). This tourism type is characterized by reducing environmental and cultural impacts that have not been considered by the other forms of conventional tourism (Wearing and Neil 1999). Furthermore, Krippendorf (1982)

F. Ramírez and J. Santana, *Environmental Education and Ecotourism*, SpringerBriefs in Environmental Science, https://doi.org/10.1007/978-3-030-01968-6_4

Fig. 4.1 Tourism destination in Villa de Leyva, Boyacá State, Colombia. **a** The Villa de Leyva village surrounded by mountains, **b** El Carmen convent, **c** Colonial architecture a legacy of the Spaniard intervention, **d** Holy *Ecce homo* priory founded in 1620 by the Spaniards and museum since 1958, **e** entrance with fossils covering floor, **f**, **g** detail of fossils, **h** fountain claimed to have no bottom, **i** internal detail of temple, **k** pulpit, **k**, **l** fossil remains, **m** entrance to garden, **n**, **o** Mary and Jesus. Photos by Fernando Ramírez. Reproduced with permission

Fig. 4.2 Rocky shore at Curaçao **a** detail, **b** surgeonfish, **c** labrids looking for food in the sand and **d** a bank of surgeonfish moving along the shore. Photos by Fernando Ramírez. Reproduced with permission

Fig. 4.3 The National Aviary in Cartagena, Colombia. **a** the endangered Sun Parakeet [*Aratinga solstitialis* (Linnaeus, 1758)], **b** flamingos, **c** peacock, and **d** Orinoco goose [*Neochen jubata* (Spix, 1825)]. Photos by Fernando Ramírez. Reproduced with permission

documented that alternative tourism has proposed a different approach contrary to mass conventional tourism. Tourism policies should not focus only on economic and technical necessities alone, but should emphasise the environment and consider the requirements of local people (Fennell 2008). According to Björk (2007) the alternative tourism concept has been used to group tourism forms that are more sustainable than mass tourism. This approach has been catalogued as categorical, with sustainable tourism on one side and mass tourism on the other. Thus, ecotourism is not the antithesis of mass tourism, but rather a complementary form of the industry (Björk 2007).

The term "*Ecotourism*" was coined by Hector Ceballos Lascuráin in 1983. According to Wearing and Neil (1999) in 1981 Ceballos-Lascuráin started using the Spanish term "*turismo ecológico*" to refer to forms of ecological tourism. This term later became shortened to "ecoturismo" in 1983 and was used by Ceballos-Lascuráin in discussions as president of PRONATURA, a conservation Non-Governmental Organization (NGO) and director general of SEDUE, the Mexican Ministry of Urban Development and Ecology (Wearing and Neil 1999). The first appearance of the word in the written form occurred in the March–April 1984 edition of American Birds as an advertisement for a tourist operation run by Ceballos-Lascuráin (Wearing and Neil 1999). His definition in written form appeared in the literature in 1987 in a paper entitled 'The future of ecotourismo' which was reprinted in the Mexico Journal of 27 January 1988 (Wearing and Neil 1999). Since the early 1980s, a number of researchers worldwide have provided multiple views on the field of ecotourism. Ecotourism has been considered an alternative to mass tourism, a culture-based tourism having a sustainable nature (Weaver 2001; Fennell 2008). Ecotourism has developed in response to unsustainable modes of tourism e.g., mass tourism. Ecotourism proposes a more sustainable and nature based form of tourism seeking to make fundamental links between biodiversity and cultural backgrounds. A careful examination of the interpretations of ecotourism shows that the concept of ecotourism seems to be the solution to all tourism problems such as environmental conservation, economic development, cultural preservation, poverty alleviation, etc. (Cobbinah 2015).

Though ecotourism offers a sustainable option for obtaining resources, it should help educate the public as well, in order to foster nature conservancy. An innovative way of merging education and ecotourism is carried out by the University of Navarra in Spain in their degree programme in Environmental Sciences. Through their *Environment and Landscapes Programme*, students are able to study ecosystems in various parts of the world. The programme is included in the curriculum and in the cost of the tuition and includes local, national, and international visits. The objective of the programme is "is to promote a deep scientific and cultural understanding of the inter-relationship between humans and the earth" (Universidad de Navarra n/d). Local or regional visits are carried out during one day. National visits are carried out over a weekend, and international visits include trips to Wales, California, Costa Rica, and Ecuador. These are carried out during term breaks. In all, the students study a total of 360 landscapes in 120 days in the field, over their four years of college. These field trips are not considered add-ons, nor extracurricular activities. Rather, they are considered a learning methodology, as students put into practice what they have seen in the classroom (Universidad de Navarra n/d). Students visit sites which have suffered environmental degradation, as well as sites which are protected, leading to a well-rounded and meaningful knowledge of how conservation practices actually work. For environmental education to be effective, hands-on contact with nature is essential.

References

Arendt H (2012) La condición humana. Paidos, Barcelona

Björk P (2007) Definition paradoxes: from concept to definition. In: Higham J (ed) Critical issues in ecotourism: understanding a complex tourism phenomenon. Elsevier, Amsterdam, pp 23–45

Cobbinah PB (2015) Contextualising the meaning of ecotourism. Tourism Manage Pers 16:179–189

Fennell D (2008) Ecotourism, 3rd edn. Routledge, New York

Krippendorf J (1982) Towards new tourism policies. Tourism Manage 3:135–148

Wearing S, Neil J (1999) Ecotourism: impacts potentials and possibilities. Butterworth-Heinemann, Oxford

Weaver D (2001) Ecotourism. Wiley, Milton

Universidad de Navarra (n/d) Environment and landscapes program. Retrieved from https://www.unav.edu/web/environment-landscapes-program/brief-information

Chapter 5
Philosophy of Ecotourism

The philosophy of ecotourism has been examined by several authors, including Neil (1999) and Fennell (2002). Ecotourism philosophy is dominated by Western thought, thus many valuable lessons can only be learned by deconstructing its meaning to find application in various socio-cultural contexts (Nowaczek et al. 2007). In the West, perceptions of ecotourism and national parks are influenced by historico-philosophical attitudes to wilderness and nature (Cochrane 2007).

The philosophical basis of ecotourism is one of the fundamental aspects that needs to be explored. Few investigations have focused on this particular aspect. The philosophy of ecotourism is deeply connected to nature. What is meant by living according to nature? According to Cleanthes, a Stoic philosopher, to live according to nature, is to define what good consists of and the purpose of human life (Cavallé 2006). The nature of human thinking is understood as the principle that sustains the world and as the unique law of all natural laws (Cavallé 2006). Nature, therefore, is to know how to live in conformity, harmony and wisdom as we are identified with nature. Nature is not a blind reality, but a source of light where human intelligence arises, and because human intelligence is the fabric of things, we do not need to make sense of things that already make sense, the unique sense (Cavallé 2006). Life manifests in life, this is when we understand, and it all begins. This is, life on our planet is unfathomable, unlimited, and permanent real essence of all that exists, at least for us humans, it is directed by a superior being understood as a supreme being that is a distinct reality from that of the world and from the world´s things. One cannot escape from the reality called life, it is impossible to live without a habitat, we know what we consider real based on life itself (Plotinus 1985).

Plotinus, a Stoic philosopher, considers that the material, phenomenal world, or Nature, is understood not as a realm of reality in its own right, but as the external and derivative aspect of an ideal world constituted by the generative power of a transcendent principle of Unity, Being and Goodness (Wildberg 2006). This power expresses itself in three different manifestations, Intellect, Soul, and Nature, in axiologically and aetiologically descending order. Physical matter, on which more later, appears at, and is involved in, that stratum of the ontology at which the information contained in

F. Ramírez and J. Santana, *Environmental Education and Ecotourism*, SpringerBriefs in Environmental Science, https://doi.org/10.1007/978-3-030-01968-6_5

Intellect ceases to be spiritually productive and becomes physically manifest instead, breaking up into fragmented copies and partial images scattered in space and time, like traces of rays of lights dissipating into an empty darkness, sporadically illuminating it here and there (Wildberg 2006). Furthermore, Plotinus considers that Nature is the sum total of the natural world, and every natural being in it, dead or alive (although nothing is really dead), is engaged in quiet contemplation of that which really exists, the ideal entities; furthermore, this thought process originates in Soul (which in turn is a thought of Intellect), and the natural world we are familiar with is nothing but an effortless fallout of the quiet act of Nature's contemplation (Wildberg 2006).

In the opening lines of the *Meteorology*, Aristotle outlines a programme for the investigation of the natural world (Falcon 2005). For Aristotle, the natural world is a causal system in which the direction of the explanation is from the celestial to the sublunary world only. Aristotle conceives of the study of the sublunary and the celestial world as forming a single science: the science of nature or natural science (Falcon 2005). Aristotle's science of nature is a distinctly organized investigation of the natural world. Aristotle does not think of the science of nature as a collection of loosely connected, if not disconnected, investigations. On the contrary, the investigations listed at the beginning of the *Meteorology* are distinct but related (Falcon 2005). There is no doubt that Aristotle's investigation is carefully structured: it begins with an examination of the first causes of nature and natural change in general, continues with a study of the celestial region, and ends with an investigation of the sublunary world, including a study of plants and animals (Falcon 2005). By dealing with nature and change, Aristotle's *Physics* provides a foundation for the entire investigation of the natural world. By saying that the *Physics* is concerned with the first causes of nature and change in general, Aristotle makes it clear that the *Physics* provides the explanatory resources and the principles for a sensible investigation of the natural world including plants and animals (Falcon 2005).

The idea of nature inspired by philosophy is one of the key elements for identifying a philosophical perspective in ecotourism. The philosophy of ecotourism is then identified as having several elements (Fig. 5.1). We here propose the philosophy of ecotourism is the harmonious contemplation of nature determined by one's knowledge, level of consciousness, connections and principles that exist in the natural world. When we refer to nature we include living as well as non-living things. The connections and principles refer to causes and the level of consciousness necessary for the appreciation of the natural world.

The philosophy of ecotourism is based on the idea that *unique* places are considered paradises. Inhabitants of these *unique* places, as well as visitors, should visit a tourism destination with consciousness i.e., the place should be conserved intact as originally found. Moreover, the visitor should have knowledge about the visited place contributing to its natural development by returning organisms (e.g., trees, plants, animals, etc.) that have been taken out after years of ecological destruction. This implies research-derived information on flora, fauna, general environmental issues and the development of an interdisciplinary team of biologists, ecologists, engineers, economists, sociologists, anthropologists and philosophers, etc.

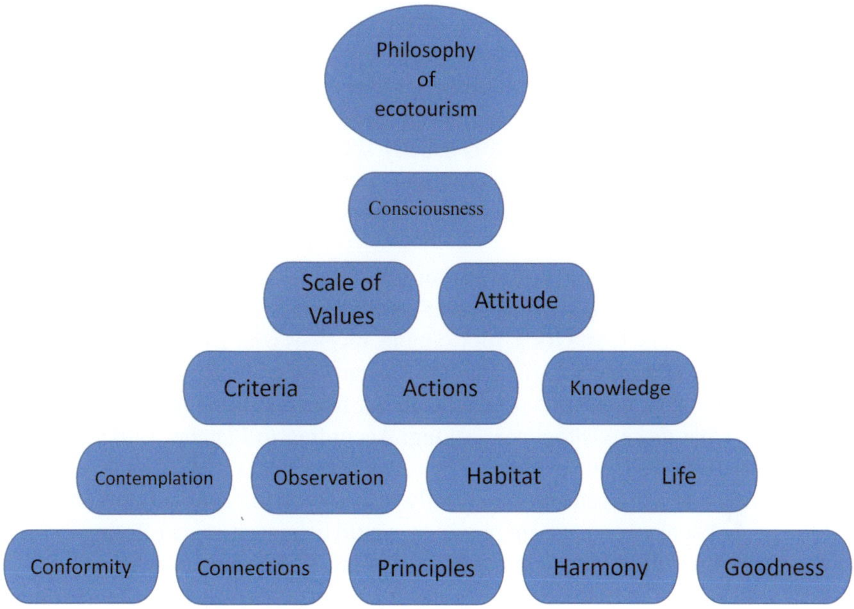

Fig. 5.1 The philosophy of ecotourism depicting important elements

One of the core ideas of the philosophy of ecotourism is the value of nature. Although, several disciplines provide valuation e.g., economists have designed valuations systems to quantify natural resources (Barbier et al. 1997), none of them provides in-depth perspectives, as philosophy does. The value of nature in philosophy can be explored by referring to the value of nature itself. However, the valuation system of philosophy is core to all disciplines because it provides the first causes and principles. The scale of values is tuned with virtues that inspire appreciation and respect for nature. The virtues are truth, beauty, caring, compassion, creativity, devotion, gratitude, humbleness, etc. The scale of values seeks to connect ecotourism with the environment and nature protection actions. The philosophy of ecotourism is key to planning and developmental ecotourism projects. To date, ecotourism is lacking a strong philosophical basis that needs to be established first at the base of definitions, objectives and perspectives in ecotourism. Core environmental values which would be articulated by ecotourists over other types of tourists, as being fundamental to the travel experience (Fennell 2002). These may include reducing/reusing/recycling, harmony, exploration, multiculturalism, preservation of landscapes, biodiversity conservation, integrity, learning, service and knowledge. In regards to this last item, knowledge is an important value because information changes at an incredible rate (Fennell 2002). Ecotourism service providers and ecotourists share many of the same values—an essential factor for successful ecotrips. From the organization's standpoint, such values must be freely communicated through all levels of the firm, but also to ecotourists as well as members of the community, who might also benefit from the positive environmental values of the service provider (Fennell 2002).

Environmental education is an important aspect that needs to be linked to the philosophy of ecotourism. It is common practice to focus on physical aspects of the environment, because this topic is usually studied as part of the curriculum of biology, or life sciences, or geography. A more integrated view of nature would also include moral and ethical issues, political and social aspects, cultural and historical viewpoints of nature.

Logically, ecotourists are visitors from urban locations; a person who lives in the rainforest has no need to visit one as an ecotourist. A large number of these tourists are foreigners. Costa Rica, for example, receives over 2 million international visitors per year (Barquero 2016; Instituto Costarricense de Turismo 2008). From an economic point of view, this is beneficial, but from a cultural point of view, it is valid to ask what these visitors are seeking. According to Fletcher (2015, p. 342), "… ecotourists desire intense, physical, visceral experiences that give them a sense of completion and achievement—especially those involving (limited) hardship and suffering…". Yet most ecotourism providers would believe they are offering tourists an opportunity to fall in love with nature, and commit to its preservation.

For Fletcher (2015), both ecotourists and environmental education take a Western perspective. A more complete environmental education programme must attempt to consider the views and needs of the local population as well.

References

Barbier EB, Acreman M, Knowler D (1997) Economic valuation of wetlands. Ramsar Convention Bureau, Gland

Barquero M (2016) País logra récords en divisas y visitantes por el turismo. La Nación. Available via: https://www.nacion.com/economia/indicadores/pais-logra-records-en-divisas-y-visitantes-por-el-turismo/G7SYUNFXRBASRMHYFBKRVIFOUE/story/. Accessed 1 Aug 2018

Cavallé M (2006) La sabiduría recobrada. Ediciones Martínez Roca, Madrid

Cochrane J (2007) Ecotourism and biodiversity conservation in Asia: institutional challenges and opportunities. In: Higham J (ed) Critical issues in ecotourism: understanding a complex tourism phenomenon. Elsevier, Amsterdam, pp 287–307

Falcon A (2005) Aristotle and the science of nature: unity without uniformity. Cambridge University Press, Cambridge

Fennell DA (2002) Ecotourism programme planning. CABI Publishing

Fletcher R (2015) Nature is a nice place to save but I wouldn't want to live there: environmental education and the ecotourist gaze. Environ Educ Res 21(3):338–350. https://doi.org/10.1080/13504622.2014.993930

Instituto de Turismo Costarricense (2008) Anuario Estadístico. Available at: http://www.bncr.fi.cr/bn/turismo/dowlands/BNCR-AnuariodeTurismo2008.pdf. Accessed 1 Aug 2018

Neil J (1999) If ecotourism is not just an activity but a philosophy, which philosophy? In: Wearing S, Neil J (eds) Ecotourism: impacts, potentials and possibilities. Reed Educational and Professional Publishing Ltd., Boston

Nowaczek A, Moran-Cahusac C, Fennell DA (2007) Against the current: striving for ethical ecotourism. In: Higham J (ed) Critical issues in ecotourism: understanding a complex tourism phenomenon. Elsevier, Amsterdam

Plotinus (1985) Enéadas: Libros III y IV. Editorial Gredos, Madrid

Wildberg C (2006) A world of thoughts: Plotinus on nature and contemplation (ENN. III.8 [30] 1–6). In: Chiaradonna R, Trabattoni F (eds) Physics and philosophy of nature in Greek Neoplatonism. Proceedings of the European science foundation exploratory workshop, Il Ciocco, Castelvecchio Pascoli, 22–24 June 2006. pp 121–143

Chapter 6
National Parks and Biodiversity Conservation

Protected areas have been considered important sites for biodiversity conservation. These areas are considered as the last sanctuaries for endangered and vulnerable species worldwide (Fig. 6.1). Protected areas includes around 15.4% of the world's land area and 3.4% of the global ocean area (UNEP 2015). The 2014 United Nations List of Protected Areas covers information of over 209,000 protected areas, from 193 countries, covering more than 32 million km^2 and distributed within 11 regions: areas beyond National jurisdiction, Africa, Asia, Caribbean, Central America, Europe, Middle East, North America, Oceania, South America and Southern Oceans (Deguignet et al. 2014). Over 30 million square kilometers have become protected in the last 50 years and the number of protected areas designated and/or recognized by countries has doubled every decade for the last 20 years (Deguignet et al. 2014). According to the IUCN (2008) definition, a protected area is considered "a clearly defined geographical space, recognized, dedicated and managed, through legal or other effective means, to achieve the long term conservation of nature with associated ecosystem services and cultural values". With their emphasis on preserving the natural environment, protected areas have obvious appeal to the ecotourism sector, which is based primarily on natural attractions (Lawton 2001). National parks are regarded as places which provide opportunities for physical, mental and spiritual refreshment and also preserve elements of a wilderness which, perhaps more in imagination than in reality, provides a contrast and counterbalance to the stable, ordered world of towns and managed countryside (Cochrane 2007). Protected areas provide opportunities to develop ecotourism. However, possible ecotourism development and projects in protected areas need to be defined under the criteria for conservation and restoration. These criteria establish that first, it is necessary to determine if it is best to totally protect-meaning no establishment of ecotourism—or to use wisely.

F. Ramírez and J. Santana, *Environmental Education and Ecotourism*, SpringerBriefs in Environmental Science, https://doi.org/10.1007/978-3-030-01968-6_6

Fig. 6.1 Santuario de Flora Isla Corota, Nariño State, Colombia—a Ramsar Site. **a** entrance, **b** main path, **c** La Cocha Lagoon, **d** detail of bromeliads and **e** trees with bromeliads. Note that this protected area is an Island within La Cocha Lagoon. Visitors have to travel by boat to reach the island. Photos by Fernando Ramírez. Reproduced with Permission

The first recognized protected areas were established in the 19th century. Yellowstone, established in 1870, is widely regarded as the world's first national park, and since then, the overall number of national parks and other protected areas has proliferated (Lawton 2001). The International Union for Conservation of Nature (IUCN) has established a classification system for protected areas worldwide (Table 6.1). The categories are recognized by international organizations such as the United Nations and by many national governments as the global standard for defining and recording protected areas and as such are increasingly being incorporated into government legislation (IUCN 2014). The IUCN includes seven protected areas categories numbered 1a to VI (Table 6.1). Conventional tourism activities become more compati-

Table 6.1 IUCN protected areas categories system

Category	Designation	Description
1a	Strict nature reserve	Strictly protected areas set aside to protect biodiversity and also possibly geological/geomorphical features, where human visitation, use and impacts are strictly controlled and limited to ensure protection of the conservation values. Such protected areas can serve as indispensable reference areas for scientific research and monitoring
1b	Wilderness area	These areas are usually large unmodified or slightly modified areas, retaining their natural character and influence without permanent or significant human habitation, which are protected and managed so as to preserve their natural condition
II	National park	These areas are large natural or near natural areas set aside to protect large-scale ecological processes, along with the complement of species and ecosystems characteristic of the area, which also provide a foundation for environmentally and culturally compatible, spiritual, scientific, educational, recreational, and visitor opportunities
III	Natural monument or feature	These areas are set aside to protect a specific natural monument, which can be a landform, sea mount, submarine cavern, geological feature such as a cave or even a living feature such as an ancient grove. They are generally quite small protected areas and often have high visitor value
IV	Habitat/species management area	These areas aim to protect particular species or habitats and management reflects this priority. Many Category IV protected areas will need regular, active interventions to address the requirements of particular species or to maintain habitats, but this is not a requirement of the category
V	Protected landscape/seascape	A protected area where the interaction of people and nature over time has produced an area of distinct character with significant, ecological, biological, cultural and scenic value: and where safeguarding the integrity of this interaction is vital to protecting and sustaining the area and its associated nature conservation and other values
VI	Protected area with sustainable use of natural resources	These areas conserve ecosystems and habitats together with associated cultural values and traditional natural resource management systems. They are generally large, with most of the area in a natural condition, where a proportion is under sustainable natural resource management and where low-level non-industrial use of natural resources compatible with nature conservation is seen as one of the main aims of the area

Source IUCN 2014

ble in the higher numbered categories (Lawton 2001). The status of ecotourism is more complex, with its soft and hard manifestations displaying very different trends. Soft ecotourism is incompatible with Category I areas, but highly compatible with Categories II and III (Lawton 2001). For the remaining categories, the compatibility is reduced, but still high, given the nature of this type of ecotourism. Hard ecotourism has a qualified place in Category I and, like its soft counterpart, displays high compatibility in Categories II and III (Lawton 2001). Category I protected areas, such as strict biological reserves, with their strict prohibitions on human activity, accommodate at best a small number of 'hard' ecotourism activities. These are likely to entail scientific and/or educational activities (Lawton 2001). In contrast, the national parks of Category II and, to a lesser extent, Category III protected areas, are highly compatible with ecotourism, and dominate the empirical literature as high profile ecotourism venues (Lawton 2001). The designation of a protected area for ecotourism should meet the following components (1) entrance points with maps and signage with attractions and services provided by the park, (2) walkways that communicate the various areas within the park, (3) infrastructure, e.g. restrooms, bird watching towers, information sites, etc., (4) medical facilities, and (5) guides and patrol guards with communication systems.

Zoning National Parks and protected areas is another important aspect that needs to be identified for a consistent ecotourism programme. Park zones are established on the basis of natural resources as well as the need to absorb recreational use (Fennell 2008). Protected area zoning is a decision-making issue that inherently requires the evaluation of multiple land attributes according to multiple objectives (Geneletti and Duren 2008). The development of an ecotourism programme in National Parks or protected areas requires a baseline study to determine the possible impacts and the best area for infrastructure. This should also consider classifying attractions within the National Park or protected area. This classification requires enumerating following aspects:

Abiotic aspects (non-living)

(a) Geographic location: Coordinates and location on a map
(b) Orography: hills, mountains, plains, etc.
(c) Climate and weather: precipitation, temperature, relative humidity, seasonality, etc.
(d) Waterways: rivers, streams, ponds, lakes, marshes, wetlands, etc.

Biotic aspects

(a) Flora: types of plants, trees, vines, shrubs, and other plant forms within the area of interest
(b) Fauna: animals of interest. Includes endemic species.

Cultural aspects

(a) Description of ethnicity within the area of interest
(b) Music: songs, dances, etc.
(c) Native foods: describes foods and beverages

(d) Rituals: describes rituals and other aspect of the target culture.

Infrastructure

(a) Buildings: includes al housing facilities
(b) Entry points: informs about the main entrance pints by air, land and waterways. Lists the number of tracks
(c) Connectivity. Internet facilities, connection to servers, information
(d) Signage: includes all information for visitors.

A more romantic view of wilderness has been developed in response to technological and industrialized transformation of Britain and Europe (Fennell 2008). To date, a number of people worldwide seek contact with nature as a form of spirituality. Spirituality has been associated with romanticism. According to Short (1991) Romanticism embodied a deeper spirituality and awareness that a simpler life was attainable without the complications of a society blemished by materialism, and could be accomplished under the following conditions: (1) untouched spaces had the greatest significance; (2) these spaces had a purity which human contact degrades; (3) wilderness was a place of deep spiritual significance; and (4) the conquest of nature was a fall from grace (Short 1991).

To date, many people from different religious backgrounds consider nature as spiritual endeavor and for deep reflection. Nature settings help restore vital aspects of a person's life. Spiritual messages i.e. those called by Christ were given in open natural spaces. National Parks and protected areas provide opportunities for people seeking spirituality by:

(1) A unique natural setting
(2) Connection of one's personal beliefs and experiences with the natural setting
(3) Reflection
(4) Meditation
(5) Spiritual change
(6) Connection to Nature/God.

References

Cochrane J (2007) Ecotourism and biodiversity conservation in Asia: institutional challenges and opportunities. In: Higham J (ed) Critical issues in ecotourism: understanding a complex tourism phenomenon. Elsevier, Amsterdam, pp 287–307

Deguignet M, Juffe-Bignoli D, Harrison J, MacSharry B, Burgess N, Kingston N (2014) United Nations list of protected areas. UNEP-WCMC, Cambridge

Fennell D (2008) Ecotourism, 3rd edn. Routledge, New York

Geneletti D, Duren I (2008) Protected area zoning for conservation and use: a combination of spatial multicriteria and multiobjective evaluation. Landscape Urban Plan 85:97–110

IUCN (2008) What is a protected area? http://www.iucn.org/about/work/programmes/gpap_hom146e/pas_gpap/

IUCN (2014) IUCN protected areas categories system. http://www.iucn.org/about/work/programmes/gpap_home/gpap_quality/gpap_pacategories/

Lawton LJ (2001) Public protected areas. In: Weaver D (ed) Encyclopedia of ecotourism. CABI Publishing, Wallingford, p 287

Short JR (1991) Imagined country. Routledge, Chapman & Hall, New York

UNEP (2015) Mapping the world's special places. http://www.unep-wcmc.org/featured-projects/mapping-the-worlds-special-places

Chapter 7
Indigenous Ecotourism

Native cultures can benefit from ecotourism. However, this can be controversial and is often dependent of the type of target culture. Some cultural groups feel that Western civilization is often invasive, threatening their beliefs, and ways of life. For example, the Indigenous people from Sierra Nevada de Santa Marta in Colombia feel that Westerners are invasive and avoid contact with *blancos* white men (Ramírez personal observation). On the other hand, other cultures are open to western thought, and feel cultural exchange is fundamental for integrating culture on a global context.

Indigenous ecotourism ventures in the Pacific Islands, Latin America, Africa and South East Asia illustrate how Indigenous groups are conserving natural areas and educating visitors while developing and controlling ecotourism on Indigenous lands and territories (Zeppel 2006). Indigenous ecotourism provides an alternative to extractive land uses such as hunting, farming, logging or mining, and it involves Indigenous people in managing tourism, culture and their own environment (Zeppel 2006). Indigenous tourism ventures are largely a response to the spread of tourism into remote and marginal areas, including national parks, nature reserves and tribal territories that are traditional living areas for many Indigenous groups (Zeppel 2006). Indigenous cultures and lands are frequently the main attraction for ecotours visiting wild and scenic natural regions such as the Amazon, Borneo, Yunnan, East Africa and Oceania (Zeppel 2006). Although, ecotourism is not considered a panacea by many authors, it can provide solutions for developing countries where poverty and resource allocation is minimal. Community ecotourism is considered a field of tourism dealing with local and Indigenous communities interacting for economic, biological and cultural benefits. In this type of tourism, not only does the community benefit, but the environment is considered an important aspect. Key aspects of Indigenous ecotourism include a nature-based product, ownership and the presentation of environmental and cultural knowledge (Zeppel 2006). Moreover, Indigenous tourism refers to tourism activity in which Indigenous people are directly involved either through *control* and/or by having their *culture* serve as the essence of the attraction (Hinch and Butler 1996). Indigenous tourism evolves when Indigenous people operate tours and cultural centers, provide visitor facilities and control tourist access to cultural

F. Ramírez and J. Santana, *Environmental Education and Ecotourism*, SpringerBriefs in Environmental Science, https://doi.org/10.1007/978-3-030-01968-6_7

sites, natural resources and tribal lands (Zeppel 2006). Acquiring western thought requires a Western view on economic transactions i.e., demand and offer of goods and services. This Western view is often controversial because it immerses Indigenous communities into inter-cultural exchanges that are considered invasive. As a consequence of this exchange, the recipient Indigenous cultures lose their identity often preferring Western amenities to their local cultural background and human history.

Ecotourism can capture biodiversity values and provide incentives for conservation, and many integrated conservation and development projects include an ecotourism component (Ohl-Schacherer et al., 2008). Ecotourism is seen as one main way for Indigenous groups to conserve and benefit from biodiversity on their traditional lands (Butcher, 2003). Indigenous groups have a profound knowledge on biodiversity related aspects. Many groups know the exact timing for planting and harvesting native plants, medicinal applications, and use nature wisely. According to Coria and Calfucura (2012), several arguments suggest that the development of Indigenous communities is compatible with ecotourism. First, there is a significant overlap between ecotourism and the development of Indigenous communities in the sense that they coexist (Coria and Calfucura 2012). Second, ecotourists generally have an explicit desire to have a positive impact, and third Indigenous communities tend to see themselves as being one with the land rather than apart from it (Coria and Calfucura 2012). Any ecotourism programme comprising Indigenous people should meet a number of points. The economic benefits of ecotourism must be accessible to the target population (Coria and Calfucura 2012). Indigenous communities need secure land tenure over the area in which ecotourism takes place. Ecotourism must promote deeper social and political justice goals to local communities, as well as the capability to make land use decisions for that area (Coria and Calfucura 2012). With greater legal recognition and control over homeland areas, culture and resources, Indigenous groups in many areas are determining appropriate types of ecotourism development in traditional lands and protected areas (Zeppel 2006). Indigenous peoples are also increasingly the owners, managers, joint venture partners or staff of ecotourism ventures, cultural sites and other tourist facilities (Zeppel 2006).

Aside from the view that Indigenous ecotourism is environmentally sound, other perspectives propose that this type of ecotourism might have negative consequences. One of these views examines that entire socio-cultural environment for Indigenous communities is strongly tied to consumptive activities (Meletis and Campbell 2007), which impact the environment having negative effects on biodiversity (Kiss 2004). Some Indigenous societies have found it difficult to manage scarce resources in a sustainable way and over-utilization of natural resources has been the norm (Fennell 2008). For example, the rate of forest clearing in some Indigenous reserves in Amazonia is alarmingly high; in one extreme case, 11.3% of a reserve was cleared in a two-year period (Fearnside 2005). Likewise, in the Colombian Amazon, the Tikunnas living near Puerto Nariño are clearing more forest to make more space for housing and crops, which alters the natural balance of forest versus agricultural and homing frontier (Ramírez personal observation). Most Indigenous peoples in this part of the world have Western based economies, culture and habits. The main income source is fishing. There is a strong fishing pressure and to date, there are a number of bans for fishing species such as the Pirarucú [*Arapaima gigas* (Schinz, 1822)]. However, there is a strong regional fishing pressure on other fish species, which is likely to impact fish populations causing an environmental fishing collapse if the appropriate measures are not taken. Fishing for ornamental species is another extractive activity that is practiced by Indigenous Peoples. Ornamental fishes from Colombia are highly priced in European and North American markets. Due to the difficult economic conditions experienced in the region, lack of governmental aid and projects, Indigenous peoples are often forced to extract and use natural resources at an alarming rate. Aside from fishing and the ornamental trade, Indigenous communities have developed tourism-based and ecotourism revenues at Puerto Nariño, at the confluence of Rio Negro and Rio Amacayacu in the Colombian Amazon. Indigenous peoples offer tours to observe birds, fishes, forest trees, and venture into floorplan lakes such as the Tarapoto Lake within the Rio Negro (Fig. 7.1). These tours are offered throughout the year, but most visitors arrive in July–August, October and December. This leaves tour guides with high season visitation three times per year and without any economic support other than the extractive fish ornamental trade and fishing during the rest of the year.

Boyd (2012) emphasizes the role of education in ecotourism, believing an activity cannot rightly be labeled ecotourism if it does not educate in some way. This education is not only in terms of content, i.e. learning about the native species, rather, it must include appropriate codes of behaviour, in line with the ethical values underpinning ecotourism. In line with these values, Boyd (2012) also establishes the need for an Indigenous stewardship of the resources. Honey (2008), as well, believes that people will protect those resources from which they receive a benefit. Thus, for ecotourism to be a successful tool for sustainable development, economic and political control must be placed in the hands of the local communities and villages—the stakeholders. For Boyd (2012), local control is still lacking in many ecotourism projects. Education of the local population is key in this sense.

Fig. 7.1 Ticuna community along the Amazon River near Leticia, Colombia. Ticunas perform a dance and sell souvenirs (left photo) and Maloca- ancestral house (right photo). Photos by Fernando Ramírez. Reproduced with permission

Torres and Athié (2011) carried out an innovative project to teach basic managerial skills to members of the Purepecha community in the town of Angahuan, in the state of Michoacán, Mexico. The town is near the site of a volcanic eruption which took place in 1943. Lava buried the nearby town of San Juan, leaving only the church spires uncovered. Since then, San Juan has become a tourist attraction and eco-tours leave on horseback from Angahuan, which is also known for its handcrafts, especially handwoven shawls.

The underlying principle for the course was the belief that managerial skills would help the participants administer their small businesses, making them more efficient, but without interfering with the local traditions or customs. There were fourteen participants—eight male and six female—all of them from the local community and all of them bilingual in Purepecha and Spanish, and with different levels of schooling. Only one of the participants had university studies, and he dropped out of the course. Nine of the participants finished the five modules and received their diplomas.

The course was offered on Saturdays, for five months. The topics included introduction to administration, basic accounting, purchasing and inventories, marketing, and taxes. The most valued were the modules on accounting and marketing, because the participants felt they had an immediate impact on their work. For example, one participant, who operated a handcrafts shop, mentioned learning the value of differentiating her products from those of other shops selling similar items. The baker mentioned the value of offering a greater variety of products. All participants felt that the course had a positive impact on their small businesses.

Though this project does not fit precisely into the category of Indigenous ecotourism, it does show the value of educating the local population to take advantage of the benefits of tourism, without sacrificing their customs or traditions.

References

Boyd SW (2012) Ecotourism: a maturing discourse with some established controversies. In: Moufakkir O, Burns P (eds) Controversies in tourism. CABI, Oxford, pp 99–114

Butcher J (2003) New moral tourism, the third world and development. The moralisation of tourism: Sun, sand … and saving the world?. Routledge, London, pp 113–136

Coria J, Calfucura E (2012) Ecotourism and the development of indigenous communities: the good, the bad, and the ugly. Ecol Econ 73:47–55

Fearnside PM (2005) Indigenous peoples as providers of environmental services in Amazonia: warning signs from Mato Grosso. In: Hall A (ed) Global impact, local action: new environmental policy in Latin America. University of London, School of Advanced Studies, Institute for the Study of the Americas, London, U.K., pp 187–198

Fennell D (2008) Ecotourism, 3rd edn. Routledge, New York

Hinch T, Butler R (1996) Indigenous tourism: a common ground for discussion. In: Butler R, Hinch T (eds) Tourism and indigenous peoples. International Thomson Business Press, London, pp 3–19

Honey M (2008) Ecotourism and sustainable development: who owns paradise?, 2nd edn. Island Press, Washington, DC

Kiss A (2004) Is community-based ecotourism a good use of biodiversity conservation funding? Trends Ecol Evol 19:232–237

Meletis ZA, Campbell LM (2007) Call it consumption! Re-conceptualizing ecotourism as consumption and consumptive. Geog Compass 1:850–870

Ohl-Schacherer J, Mannigel E, Kirkby CS, Shepard Jr GH, Yu DW (2008) Indigenous ecotourism in the Amazon: a case study of 'Casa Matsiguenka' in Manu National Park, Peru. Env Cons https://doi.org/10.1017/s0376892908004517

Torres MG, Athié R (2011) Desarrollo de habilidades empresariales en microempresarios rurales de Angahuan, Michoacán, México: Un caso de estudio. In: Paper presented at the 10th international business & economy conference, Guadalajara Mexico 6–9 January 2011

Zeppel H (2006) Indigenous ecotourism: sustainable development and management. CAB International, Wallingford, UK

Chapter 8
Environmental Impact

Ecotourism represents one of the more eco-friendly alternatives for the economic use of natural resources (Li 2004). The core basis of the field of ecotourism is deeply connected to environmental issues. Other tourism forms are less involved with caring for the environment. Although, the ecotourism sector has principles that pertain to nature, such as environmental conservation, a number of impacts have been generated as a result of human activities in nature-based destinations.

Ecotourism has proven to be a successful environmental conservation tool in many cases around the world (Das and Chatterjee 2015). This is the case of the Galapagos Islands, Costa Rica's ecotourism spots, Chitwan National Park (Nepal), Sunderbans (India), Periyar Tiger Reserve (India), Kilum-Ijim National Park (Central Africa), Cuyabeno Wildlife Reserve (Ecuadorian Amazon region), community based tourism in Indonesia, etc. (Das and Chatterjee 2015). These places reflect properly organized ecotourism, which enables local people to augment their livelihood security through employment in ecotourism related activities and small enterprise development (Das and Chatterjee 2015). Moreover, ecotourism has provided economic empowerment to local communities, increased living standards for locals, respect for their own culture, external recognition, appreciation of their resources, morale boosting, giving rise to psychological well-being (Das and Chatterjee 2015). For example, ecotourism has provided the tangible aspect of conservation as it has been helping save animals and fragile ecosystems (Libosada 2009). Moreover, ecotourism has been successful in conservation policy for its capacity to conserve large mammals' diversity in Upper Amazonia through sustainable source of income to the inhabitants (Salvador et al. 2011). Other positive views about ecotourism focus on Indigenous cultures and how these cultures have positively gained from the ecotourism experience.

The development of ecotourism destinations inevitably brings with it a number of associated impacts (Higham and Lück 2007). Critics argue that ecotourism promotes development and thus, the destruction of natural resources (McLaren 2003). Weaver (2002) notes that the environmental impacts of ecotourism can be either deliberate or inadvertent. Deliberate costs are, for example, those costs that occur with the construction of an ecolodge. Das and Chatterjee (2015), proposed that numerous studies

F. Ramírez and J. Santana, *Environmental Education and Ecotourism*, SpringerBriefs in Environmental Science, https://doi.org/10.1007/978-3-030-01968-6_8

claim that ecotourism is not very effective in promoting conservation of biodiversity. Many authors consider ecotourism as an instrument for revenue generation. The word 'ecotourism' is used to attract customers, and thereby generate more income. Many protected areas in developing countries are poorly planned, with the infrastructure and management inadequate, or even unsuitable, for ecotourism (Das and Chatterjee 2015). This is the particular case of Bogotá's urban wetlands located in Colombia, South America. Individually, they may appear to be small, but they form a strong network of wetlands when taken collectively (Ramírez et al. 2013). Today, the total wetland area within the city is about 500 ha (Pinilla 2010) and includes 15 different wetlands: Santa María de Lago, Córdoba, Juan Amarillo, Capellanía, La Vaca, El Burro, Techo, Tibanica, Jaboque, Guaymaral-Torca, La Conejera, Meandro del Say, La Isla, El Tunjo, and El Salitre (Ramírez and Fennell 2014; Ramírez and Santana 2018). Bogotá's wetlands are home to many species of terrestrial and aquatic vertebrates, invertebrates, plants, and other organisms. Many of these species are endemic, migratory, or are vulnerable, endangered, or have a data-deficient status according to the International Union for Conservation of Nature (Ramírez and Fennell 2014). However, most of these wetlands face serious threats such as (1) severe degradation because of a series of governance issues; (2) Bogota's city office (SDA), Cundinamarca state office (Corporación Autónoma Regional de Cundinamarca, CAR), and Bogotá's Water Company (Empresa de Acueducto y Alcantarillado de Bogotá, EAAB) have not made efforts to restore or preserve wetlands; (3) security issues that are barriers to further development of the ecotourism industry; (4) lack of interest by the local community in most wetlands; (5) anti-corruption measures that enable governmental funds to reach wetland management plans; and (6) involvement of the private sector and NGOs (Ramírez and Fennell 2014; Ramírez 2015).

According to Das and Chatterjee (2015), the reasons behind the problems associated with ecotourism are numerous, such as (1) revenue leakages, as labor is drawn from the urban sector instead of focusing on training of unskilled or less-skilled locals, (2) inequitable distribution of income among the locals, (3) compulsory displacement for the creation of national parks leading to large-scale loss of land, homelessness, food insecurity, loss of lives and increase in morbidity, (4) restrictions in accessing sanctuary resulting in joblessness and (5) damage to crops and livestock by wildlife, (6) sticking only to gun-and-guard approaches for preservation, (7) crowding, crime, begging, prostitution, etc. associated with the increase in the number of tourists in ecotourism spots, (8) insensitive attitude of the tourist, (9) lack of education for visitors as well as locals and (10) policy gap in terms of poor planning and improper and unethical management of ecotourism (Das and Chatterjee 2015).

Ecotourism activities have been documented to conflict with the protection of Natural Protected Areas (Polgar and Jaafar 2018). Visitors cause several impacts to the environment, e.g. wetlands. However, these impacts can be minimized by specialized infrastructure. For example, visitor trampling can be reduced by building boardwalk trail facilities. These trails allow visitors to observe wetland plants and animals without causing impact to the wetland (Polgar and Jaafar 2018). Moreover, these trails can provide essential instruction about the relevance of wetlands (Haslam et al. 2009). Other facilities that aid in the educational-ecotourism experience are water-proof bunkers with potholes for observation of mud creatures. Furthermore, floating platforms can be located near the trail. Permanent binoculars and wildlife can be placed along the trail for public use. Their use can reduce the impact caused by visitors. Video cameras can be placed within the marsh areas to record wildlife and cause no environmental impact (Polgar and Jaafar 2018). Guided tours in boats and canoes is an option if reduced noise and fauna disturbance can be achieved. Here the guide can focus on environmental education pointing out the need to preserve the wetland, biodiversity and cultural aspects. Furthermore, specialist leaflets such as elementary and advanced, entomological, and botanical are ideal for environmental education (Haslam et al. 2009). Bird watching is another activity that can be coupled with environmental education and ecotourism. This activity should be conducted in small groups of people to reduce the impact on wetland birds (Haslam et al. 2009).

The ecotourism views—either positive or negative—are closely connected within a context. The context of developed countries is likely to be more consistent with the principles and objectives of ecotourism due to the fact that these countries have numerous nature-based sites that could be the focus of ecotourism.

Educating the public is an important aspect that minimizes the impacts of any ecotourism activity. According to Kimmel (1999) ecotourism has provided an important opportunity to promote environmental education. Educational experiences are required to provide relationships and meanings to people about the places they visit and about the content to see and do there (Tisdell and Wilson 2005). Another important aspect is to focus on environmental education and fostering of pro-conservation actions and values (Tisdell and Wilson 2005). In a case-study about turtle conservation at Mon Repos Conservation Park, Queensland, Australia, Tisdell and Wilson (2005) revealed that a great amount of environmental knowledge is acquired by visitors. The experience of observing sea turtles was positive and impacted the visitors desire and intended behaviour to protect sea turtles. This is key to linking environmental education to pro-conservation sentiments and actions (Tisdell and Wilson 2005).

In the case of Kerkini and Prespes areas in northern Greece areas of Ampelakia, Agios Panteleymonas and Tavropos in central Greece, a study revealed that individuals running small ecotourism enterprises are willing to cooperate with environmental organizations. These individuals showed a high degree of environmental activism and a knowledge about their environmental education requirements (Skanavis et al. 2004). They would like to get involved in environmental education initiatives.

Walter (2009) examined how local knowledge is engaged in environmental adult education in a community-based ecotourism project in southern Thailand. In their project, ecotourists remained as guests in local homes to learn from the local community's ecotourism guides. The information taught by guides centered on local knowledge of environmental conservation efforts, tidal and marine ecosystems, traditional livelihood activities and local culture. Ecotourists learned experientially: they watch, listen and then try out a particular skill e.g. harvesting rubber, binding the claws of cultivated lobster or crafting batik cloth (Walter 2009). Local guides relayed on community elders as a source of knowledge. Within this community developed the Koh Yao Noi Eco-Tourism Club to learn about the ways-of-life of local people and the coastal resources on which they rely for their livelihood (Walter 2009). This Club has objectives and rules for ecotourists.

Education is a core objective within any ecotourism experience. However, "education" has not been subject to refinement and scrutiny (Zanotti and Chernela 2008). What is meant by 'study' or 'education' varies depending on who is being educated, by whom and toward what end (Zanotti and Chernela 2008). In their study, Zanotti and Chernela (2008) examined an ecotourism project for university students designed by an anthropologist, an ecologist, and the Kayapó Indigenous community in Brazil. These investigators looked at the following assumptions, (1) who or what is considered a reliable source of information?; (2) what information is identified as 'educational' within the ecotourism experience?; (3) who are the learning targets?; and (4) what are the objectives of education within the context of ecotourism? (Zanotti and Chernela 2008). Their conclusions invite to review the assumptions and interpretations underlying the role of education in ecotourism. For instance, it was found by the authors that the the flow of information was not unidirectional, and that instead it was a dialogue between cultures. Education needs to be re-evaluated in the light of cultural-based knowledge and not based on Western thought. Local knowledge is core to the ecotourism experience.

References

Das M, Chatterjee B (2015) Ecotourism: a panacea or a predicament? Tourism Manag Pers 14:3–16

Haslam S, Klötzli F, Sukopp H, Szczepański A (2009) The management of wetlands. In: Westlake D, Květ J, Szczepański A (eds) The production ecology of wetlands: the IBP synthesis. Cambridge University Press, Cambridge, pp 405–464

Higham J, Lück M (2007) Ecotourism: pondering the paradoxe. In: Higham J (ed) Critical issues in ecotourism: understanding a complex tourism phenomenon. Elsevier, Amsterdam, pp 117–135

Kimmel JR (1999) Ecotourism as environmental learning. J Environ Educ 30:40–44. https://doi.org/10.1080/00958969909601869

Li W (2004) Environmental management indicators for ecotourism in China's nature reserves: a case study in Tianmushan nature reserve. Tourism Manag 25:559–564

Libosada CM Jr (2009) Business or leisure? Economic development and resource protection-concepts and practices in sustainable tourism. Ocean Coastal Manag 52:390–394

McLaren D (2003) Rethinking tourism and ecotravel, 2nd edn. Kumarian Press, Bloomfield

Pinilla G (2010) An index of limnological conditions for urban wetlands of Bogotá city. Colombia Ecol Ind 10:848–856

Polgar G, Jaafar Z (2018) Endangered forested wetlands of Sundaland: ecology, connectivity, conservation. In: Polgar G, Jaafar Z (eds) Ecotourism and the future of the forested wetlands of Sunsaland. Springer, Switzerland, pp 89–93

Ramírez F, Davenport TL, Kallarackal J (2013) Bogota's urban wetlands. In: Lavigne G, Cote C (eds) Colombia social, economic and environmental issues. Nova, New York, pp 1–80

Ramírez F, Fennell D (2014) A comprehensive framework for ecotourism and wetland restoration: the case of Bogotá, Colombia. J Ecotourism 13:128–151

Ramírez F (2015) Bogotá's urban wetlands: biodiversity, policy, threats and ecotourism. In: Advances in environmental research, vol 45. Nova Publishers, New York

Ramírez F, Santana JC (2018) Key lessons learned by teaching ecotourism to undergraduate students in Bogotá's urban wetlands. Appl Environ Educ Commun 1–18. https://doi.org/10.1080/1533015x.2018.1454359

Salvador S, Clavero M, Pitman RL (2011) Large mammal species richness and habitat use in an upper Amazonian forest used for ecotourism. Mammal Biol 76:115–123

Skanavis C, Matsinos YG, Petreniti V (2004) Environmental education potential for greek ecotourism. Int J Environ Stud 61:735–745. https://doi.org/10.1080/0020723042000271668

Tisdell C, Wilson C (2005) Perceived impacts of ecotourism on environmental learning and conservation: turtle watching as a case study. Environ Dev Sustain 7:291–302. https://doi.org/10.1007/s10668-004-7619-6

Walter P (2009) Local knowledge and adult learning in environmental adult education: community-based ecotourism in southern Thailand. Int J Lifelong Educ 28:513–532. https://doi.org/10.1080/02601370903031363

Weaver DB (2002) The evolving of ecotourism and its potential impacts. Int J Sust Dev 5:251–264

Zanotti L, Chernela J (2008) Conflicting cultures of nature: ecotourism, education and the Kayapo of the Brazilian amazon. Tour Geogr 10:495–521. https://doi.org/10.1080/14616680802434114

Chapter 9
Wetland Ecotourism

Wetlands are important ecosystems worldwide that have provided a number of ecosystems services. Ecosystem services are a wide range of social, economic, cultural, and environmental benefits that people receive from ecosystems (Costanza et al. 1997). Wetlands provide a number of ecosystem services such as maintaining water quality and supply, decreasing suspended matter, regulating atmospheric gases, sequestering carbon, regulating water regimes, retaining pollutants, sustaining unique indigenous biota, protecting shorelines, and providing recreational, cultural and educational resources (Dise 2009; Haslam et al. 2009). Also, wetlands are considered rich in biodiversity and are among the most vulnerable and sensitive ecosystems (Smardon 2009). They are important habitats for migratory birds and number of endangered reptile, mammal, amphibian, and plant species (Fig. 9.1) (Smardon 2009).

Over the last 100 years it has been estimated that 60% of the wetlands worldwide have been destroyed (Nesmith et al. 2016). According to van der Valk (2006) about 50% of the wetlands of the world have been lost and in some regions, nearly 99%. The major cause of wetland loss is due to the conversion of wetlands through human induced activities such as filling or draining into agricultural land, silviculture and/or human settlements (Fraser and Keddy 2005). Currently, wetlands are facing numerous environmental issues such as land-filling, water pollution, eutrophication, invasive species, sewage, conversion into agricultural land, agricultural run-off, global climate change, etc. (van der Valk 2006; Ramírez et al. 2013). One of the key points to preserve wetlands is environmental education. The key message is to increase awareness and make people i.e. children, teenagers, adults and the elderly more prone to conservation initiatives and to care for wetland environments. This paradigm change can be taught through a dialogue that takes place within a wetland system where people are able to observe and reflect about the values and ecosystem functions of a wetland. Other recreational activities such as photography, recording, bird watching, can be used to focus on environmental education.

The Convention on Wetlands is an intergovernmental treaty initially adopted on February 1971 at Ramsar, Iran (Ramsar 2016). This convention has emphasised

F. Ramírez and J. Santana, *Environmental Education and Ecotourism*, SpringerBriefs in Environmental Science, https://doi.org/10.1007/978-3-030-01968-6_9

Fig. 9.1 Wetland environments **a** lake near Cartagena, Colombia, **b** mangrove forest in Cancún, Mexico, **c** Lake Tarapoto, colombian Amazon, a Ramsar site and **d** Laguna de La Cocha, Nariño State, Colombia—a Ramsar site. Photos by Fernando Ramírez. Reproduced with Permission

all aspects related to wise use and conservation of wetlands in different countries. Ramsar has been developed to increase international awareness about the rate of wetland disappearance as a result of lack of knowledge of their functions, goods, values and ecosystem services (Ramsar 2016). The core Ramsar message is the request for a sustainable wetland use. To date, 2315 wetland sites worldwide have been included into the Ramsar Convention. These comprise 245,666,885 ha (Ramsar 2018). Within the Ramsar framework, several goals and priorities have been identified for wetland conservation and sustainable use. Within the context of Ramsar, a practical definition of wetland has been coined—"*wetlands are areas of marsh, fen, peatland or water, whether natural or artificial, permanent or temporary, with water that is static or flowing, fresh, brackish or salt, including areas of marine water the depth of which at low tide does not exceed six metres*" (Ramsar 2016, p. 9). It is important to define what is meant by a wetland because this makes it easier to propose fundamental aspects in conservation and, sustainable use initiatives and frameworks within the context of environmental education and ecotourism.

Ramsar sites are ideal for nature conservation, biodiversity, environmental education and ecotourism purposes. Ramsar's wise use of wetlands included carrying out conservation education and increasing public awareness (Ramsar 2016). This involved developing conservation education programs associated to wetland reserve networks. In such programs, trained personnel in education are ideal for effective conservation purposes (Ramsar 2016). Ramsar developed the "The Communication, Education, Participation, and Awareness (CEPA) program". This program encourages and mobilizes both individual and collective action for biodiversity conservation (Hesselink 2007). The CEPA program has an oversight panel, focal points per country, resources and activities, network, wetlands educational centers, culture and wetlands (CEPA program 2016). In education, it develops understanding, attitudes of concern for the environment, clarifies values, and encourages skills and motivation for environmental actions (Hesselink 2007). The CEPA program has been involved with various biodiversity educational activities in schools and universities. The CEPA program has been producing materials such as books and booklets to target specific communities and objectives. Examples of actions conducted through the CEPA program in Bolivia show conservation and education activities that have strengthening mechanisms to bring together people, information, resources, and materials experiences for collaboration, cooperation and exchange, for building a strong base for biodiversity, water, and forest action. The CEPA program has also maximised the use of local networks in India such as in the northern coastal Andhara, where it has been possible to obtain valuable micro-level information on biodiversity from remote areas (Hesselink 2007). The CEPA program has focused on the education sector in the German School-Net by a pilot activity to observe the reaction of schools, providing materials such as CDs, HTML eLearning tools, and published articles. These activities involve practical field work observations and a reporting mechanism on the Internet (Hesselink 2007).

Wetland ecotourism is a field that has attracted interest from researchers and the general public worldwide. Ecotourism can be viewed as effective for wetland environmental education. Within this context it is important to generate frameworks for ecotourism and wetland restoration (Ramírez and Fennell 2014). Ecotourism is proposed as a land-use option in the context of its value as a tool for restoration (Ramírez and Fennell 2014). These authors proposed within the environmental dimension, biodiversity conservation measures that include educational programs involving the local community. Educational programs could help people to recognise the urgency and vital significance of wetland protection (Wang et al. 2008). Environmental education promotes behavioral changes facilitating wetland conservation and habitat protection (Ibrahim et al. 2012). Furthermore, public education awareness has been important to interact, exchange experiences and ideas among stakeholders in wetland settings (Macharia et al. 2010). In the case of the highland wetlands in central Kenya, public education awareness was enacted through a series of seminars carried out in different phases. In phase one and two, the community identifies the importance of wetlands and their environmental problems. In phase three, alternative sustainable use practices were discussed (Macharia et al. 2010). In this scenario, the community identified ecotourism as one of the most reliable and environmentally friendly activities and recognized the biodiversity rich opportunities of wetlands as well as the abundant biodiversity found within swamps.

The link between environmental education and ecotourism in wetland settings has been analysed by Ramírez and Santana (2018) in three wetland systems in Bogotá,

Fig. 9.2 Bogotá's wetlands in Colombia. **a** Córdoba wetland, **b** Guaymaral wetland and **c–d** Santa María del Lago wetland. Photos by Fernando Ramírez. Reproduced with permission

Colombia (Fig. 9.2). The aim of their work was to describe key lessons that can be learned by teaching ecotourism in Santa María del Lago, Córdoba, and Guaymaral wetlands. Ramírez and Santana (2018) analysed and developed course content based on seven key components related to wetlands: (1) infrastructure for ecotourism, (2) biodiversity, (3) cultural aspects, e.g., wetland history, (4) wetland management issues, (5) environmental education, (6) conservation and restoration, and (7) community involvement. Within each component, key lessons were developed where biodiversity aspects are a key point for the ecotourism class. In environmental education they proposed key lessons for the implementation of a comprehensive ecotourism educational experience based on a holistic approach for deep understanding, and transformation of behavior, indicating wetland environmental problems and suggesting possible solutions. Observation of wetland plants and animals connects to students´ prior knowledge. On-site experience contributes to the pro-environmental learning outcomes. It should be pointed out that most of Bogotá´s wetlands have been forgotten by the Government and that few actions have been carried out improve their conservation and environmental issues (Ramírez and Fennell 2014).

Environmental education has been practiced in different settings, among them nature centers, zoos and museums (Winther et al. 2010). These environments serve as outdoor learning labs that give students the opportunity to generate awareness and interact with the local environment (Winther et al. 2010). This outdoor experience can also be applied to a wetland setting. Thus, students have opportunities to obtain knowledge through exploratory, self-guided, and assessment-free experiences (Winther et al. 2010). This gives the learning process a multisensory dimension beyond classroom textbooks, workshops and within-classroom activities (Winther et al. 2010). In the case of wetlands, the location itself is a completely new experience for students. Also, the texture of plants or animals such as insects are new experiences. Observation of the different vegetation patterns and the smell of flowers within the wetland is also inspiring. Students grasp beauty and become immersed within an emotional and groundbreaking experience.

References

CEPA Program (2016) The RAMSAR CEPA programme. https://www.ramsar.org/activity/the-ramsar-cepa-programme

Costanza R, d'Arge R, de Groot R, Farber S, Grasso M, Hannon B, Limburg K, Naeem S, O'Neill RV, Paruelo J, Raskin RG, Sutton P, van den Belt M (1997) The value of the world's ecosystem services and natural capital. Nature 387:253–260

Dise NB (2009) Peatland response to global change. Science 326:810–811

Fraser L, Keddy P (2005) The future of large wetlands: a global perspective. In: Fraser L, Keddy P (eds) The world's largest wetlands: ecology and conservation. Cambridge University Press, Cambridge, pp 446–468

Haslam S, Klötzli F, Sukopp H, Szczepański A (2009) The management of wetlands. In: Westlake D, Květ J, Szczepański A (eds) The production ecology of wetlands: the IBP synthesis. Cambridge University Press, Cambridge, pp 405–464

Hesselink FJ (2007) Communication, education and public awareness, a toolkit for the convention on biological convention, Montreal. Design & Typesetting: Em Dash Design

Ibrahim I, Aminudin N, Yahya MAY, Suleiman AIY (2012) Education for wetlands: public perception in Malaysia. Procedia—Soc Behav Sci 42:159–165. https://doi.org/10.1016/j.sbspro.2012.04.177

Macharia JM, Thenya T, Ndiritu GG (2010) Management of highland wetlands in central Kenya: the importance of community education, awareness and eco-tourism in biodiversity conservation. Biodiversity 11:85–90. https://doi.org/10.1080/14888386.2010.9712652

Nesmith SM, Wynveen CJ, Dixon EM et al (2016) Exploring educators' environmental education attitudes and efficacy: insights gleaned from a Texas wetland academy. Int J Sci Educ Part B Commun Public Engagem 6:303–324. https://doi.org/10.1080/21548455.2015.1078519

Ramírez F, Davenport TL, Kallarackal J (2013) Bogota´'s urban wetlands. In: Lavigne G, Cote C (eds) Colombia social, economic and environmental issues. Nova, New York, pp 1–80

Ramírez F, Fennell D (2014) A comprehensive framework for ecotourism and wetland restoration: the case of Bogotá, Colombia. J Ecotourism 13:128–151. https://doi.org/10.1080/14724049.2015.1011161

Ramírez F, Santana JC (2018) Key lessons learned by teaching ecotourism to undergraduate students in Bogotá's urban wetlands. Appl Environ Educ Commun 1–18. https://doi.org/10.1080/1533015x.2018.1454359

Ramsar (2016) An introduction to the Ramsar convention of wetlands. Sub-Series I: Handbook 1. International cooperation on Wetlands, 7th edn. (previously The Ramsar Convention Manual). Ramsar Convention Secretariat, Gland, Switzerland

Ramsar (2018) Servicio de información sobre sitios Ramsar. https://rsis.ramsar.org/es?language=es

Smardon R (2009) Sustaining the world's wetlands setting policy and resolving conflicts. Springer, New York

van der Valk A (2006) The biology of freshwater wetlands. Oxford University Press, Oxford

Wang Y, Yao Y, Ju M (2008) Wise use of wetlands: current state of protection and utilization of Chinese wetlands and recommendations for improvement. Environ Manage 41:793–808

Winther A, Sadler K, Saunders G (2010) Approaches to environmental education. In: Bodzin A, Shiner B, Weaver K (eds) The inclusion of environmental education in science teacher education. Springer, New York, pp 31–49

Chapter 10
Conclusion

This book has covered diverse aspects related to ecotourism and environmental education, among them, the role of education in biodiversity conservation. Over the last 25 years, nations around the world have recognised the importance of preserving different animal and plant species. The United Nations Organization has established biodiversity conservation as one of its major goals for the next decades, and has emphasised the importance of education in accomplishing this goal, by focusing on education for sustainable development, increasing public awareness, and training the Indigenous populations to monitor and manage natural resources. Examples from different parts of the world have shown how the goal is being accomplished.

The book has also covered diverse aspects of ecotourism, including its relevance in today's world, not only as an economic benefit for local communities, but as a source of spiritual and physical well-being for its practitioners. A third chapter discussed ecotourism as a tool in education, not only in terms of learning about nature. The acquisition of values, including appreciation for the interconnectedness of all forms of life, is also a goal of environmental education. This type of knowledge is best learned in situ. In order to educate future generations, and combat what has been termed "nature deficit disorder"—the lack of contact with nature—experiential environmental learning is crucial. This experience can be conducted in various environments e.g. wetlands.

An important chapter has dealt with the little explored philosophical underpinnings of ecotourism. The chapter recounts how nature has long been associated with spiritual, as opposed to material, aspects of human life. A love of nature is associated with ethical values, and the goals of ecotourism should include the possibility of experiencing and valuing nature, not only as a source of economic benefits, but simply because it is good in and of itself. Most philosophy of the environment, however, has its basis in Western thought. It is necessary to begin to explore how the Indigenous population views nature. The wise view of Indigenous cultures sees nature as part of the cosmos and as a sacred realm where every rock, root and animal has a function and thus deserves to be preserved. However, some Indigenous

F. Ramírez and J. Santana, *Environmental Education and Ecotourism*, SpringerBriefs in Environmental Science, https://doi.org/10.1007/978-3-030-01968-6_10

cultures influenced by Western thought, have forgotten their roots and seem unaware of the wisdom within nature. Our hope is that through consciousness, environmental education and nature-based philosophy, people awaken to conservation and environmental awareness dimensions that drive a harmonious relationship with nature and the environment.

Other chapters have dealt with the educational value of natural parks and wetlands. For environmental education to make an impact, it is necessary to remove it from the classroom and place the learning firmly in nature. Both parks and wetlands offer (usually) easily accessible areas where students can explore and learn about the environment. Youth education is an essential aspect than needs to be included in any environment and ecotourism discussion. The future generations hold the power of change and in their hands lie possible solutions for nature and biodiversity conservation. Through hands-on-activities children can get into contact with nature settings and organisms. This can be done by taking children to nature-based environments nearby e.g. wetlands, rainforests, plains, mountains, etc. Children can grasp a "real experience" first hand, with nature and connect to it through observation and their senses. Here, the teacher gives information about organisms and interesting facts about the particular environment and its importance.

Environmental education is important for ecotourism. Moreover, in the ecotourism field, several authors have mentioned the relevance of education. Around the world, several studies have been conducted linking environmental education and ecotourism. This has increased ecotourists' awareness and commitment to the environment. Ecotourists get the opportunity to live within a target community and experience first-hand the way of life of the target culture. The knowledge of the target culture is key to the ecotourism experience because it provides other views and values linked to biodiversity conservation and the environment. Particular cultures have their own environmental education values that can be transmitted to visitors. Also, their view of nature and caring for the environment is valuable as an ecotourist experience.

A common theme running through various chapters, but emphasised in Chap. 8 is the need to include the Indigenous population in different aspects of environmental education. The local population needs to be educated, not only in how to conserve their resources, but also in how to manage them, in how to educate others about them, and in how to reap the economic benefits of their resources in ways that are sustainable. However, it should be realized that some Indigenous people consider Westerners as invaders and see no benefit in ecotourism. This view is also valid and in this case, the culture should be left alone as a reminder that Western thought and activities that connect to nature are not the only consistent possibilities.

More research about environmental education and ecotourism is warranted. It is particularly important to focus on developing countries due to the diversity of cultures and biodiversity rich environments. This research can define objectives and shape the views of scientists, governments, ecotourism projects, the tourism industry, stakeholders and citizens. Some countries, like Mexico and Colombia, need more research in the field of ecotourism and environmental education to direct actions towards nature conservancy, management and policy. Education is key to wisely manage and use resources and as a mechanism to increase the consciousness of the general public.

Finally, a book about environmental education and ecotourism could not be complete without acknowledging that exploiting nature, even through sustainable projects, will have an impact on the environment in some way. It must be recognised, however, that not exploiting nature also has an impact. The goal is to minimise the negative impact of ecotourism while allowing people to enjoy the natural beauty of the world.

MIX
Papier aus verantwortungsvollen Quellen
Paper from responsible sources
FSC® C105338

If you have any concerns about our products, you can contact us on
ProductSafety@springernature.com

In case Publisher is established outside the EU, the EU authorized representative is:
Springer Nature Customer Service Center GmbH
Europaplatz 3, 69115 Heidelberg, Germany

Printed by Libri Plureos GmbH
in Hamburg, Germany